AF882733

LAUNCHING

THE

FUTURE

Insights from a Maiden Capital Campaign in Higher Education

By

Anne Okaomee, PhD

Copyright © 2026

By

Anne Okaomee, PhD

All rights reserved.

Published by:

Best Book Company

Michigan, USA

ISBN:

Paperback: 978-1-968762-42-1

Hardcover: 978-1-968762-43-8

eBook: 978-1-968762-41-4

Printed in USA

Table of Contents

Introduction

Purpose of the Book

This book explores the process of running a capital campaign, using a real-life case study to illustrate key principles and lessons. The focus is the maiden capital campaign of the University of Port Harcourt, hereafter referred to as the maiden campaign. Through this example, the book aims to provide guidance and insights for development offices embarking on similar fundraising initiatives, particularly in the context of Nigerian universities.

Importance of Capital Campaigns in University Advancement

Capital campaigns are critical instruments in university advancement. They allow institutions to mobilize significant resources for infrastructure, scholarships, research, and other strategic priorities. Beyond the immediate financial benefits, campaigns help foster a culture of giving, strengthen alumni engagement, and elevate the university's profile both locally and internationally. Understanding the mechanisms of a successful campaign is therefore vital for any university seeking sustainable growth.

Overview of the Maiden Campaign

The maiden campaign provides a rich case study for understanding the complexities and opportunities of capital fundraising. This book begins with an overview of major fundraising programs, then defines what constitutes a capital campaign. It continues with a discussion of the phases of a campaign and examines how each phase was implemented during the University of Port Harcourt's initiative. Drawing on the framework outlined in the fourth edition of *Achieving Excellence in Fundraising*, the book also highlights the lessons learned from the maiden campaign and recommends best practices for other development offices in Nigeria embarking on similar efforts.

Through this narrative, readers will gain both practical insights and an appreciation for the strategic planning, leadership, and stakeholder engagement required to execute a successful capital campaign in a higher education setting.

Chapter 1

Overview of Major Fundraising Programs

Historical Context of Philanthropic Fundraising

Philanthropic fundraising has a rich history, dating back centuries to the efforts of both religious and secular organizations seeking to address social needs. From Booker T. Washington's personal solicitation of major gifts from wealthy capitalists such as Andrew Carnegie in 1900 for Tuskegee University (Washington, 1901), to Charles Summer Ward's 1905 major fundraising drive to fund a Young Men's Christian Association (YMCA) building (Conley, 2016), institutions have employed a variety of techniques to engage donors and secure resources. Over time, these strategies have evolved significantly. In the 21st century, fundraising has become more formalized, organized, diverse, and complex (Rosso, 2016).

Key Fundraising Programs

Modern nonprofit organizations, including educational institutions, rely on several major fundraising vehicles to mobilize support: Annual Giving, Major Giving, Planned Giving, and Capital Campaigns (Seiler, 2016a; The Fund Raising School, 2017).

- Annual Giving targets a broad donor pool and is typically used to support operating expenses. It encourages participation at all giving levels, even for modest contributions.

- Major Giving focuses on a differentiated set of donors capable of making significant contributions, often defined by the organization as "major gifts."

- Planned Giving involves more complex instruments, usually designed to yield funds in the future, such as bequests, life insurance, or charitable trusts.

- Capital Campaigns represent the most comprehensive strategy, integrating annual gifts, major gifts, planned gifts, and even endowments. These campaigns are often launched with the expectation of raising substantial sums within a defined period to support major initiatives.

Evolution of Fundraising Strategies

Organizations typically develop fundraising capacity incrementally. With limited human and financial resources, they often begin with annual giving programs, progress to major giving, and, depending on the donor base, move toward planned giving initiatives. Running a capital campaign, however, demands significantly greater resources, organizational capacity, and professional competence. Ensuring campaign success requires meticulous planning, strategic leadership, and an understanding of both donor motivations and institutional capabilities. Recognizing these requirements is critical to avoid pitfalls that could undermine the campaign's objectives.

Chapter 2

Understanding Capital Campaigns

Definition and History of Capital Campaigns

A capital campaign is an intensive, multiyear fundraising drive designed to raise funds for capital and other projects within a specified period and toward a defined goal. Historically, capital campaigns were understood to fund only capital projects, such as buildings. Over time, however, this understanding has broadened. Today, capital campaigns can raise funds for a variety of purposes, including operating expenses, endowments, and other institutional needs (Conley, 2016; Newman, 2005).

Organizations across the nonprofit sector, regardless of size or type, undertake capital campaigns. The need for a campaign often arises from challenges such as gaps in funding due to dwindling state support, the desire to construct or renovate facilities, build endowments, or support operational costs (Nehls, 2008).

The origins of the capital campaign as a fundraising strategy can be traced to 1905, when Charles Summer Ward led a drive in Chicago to fund a new YMCA building. Core features of a campaign include a specified monetary goal, a defined campaign period with a start and end date, and a clear set of projects that the raised funds will support (Conley, 2016).

Types of Capital Campaigns

Capital campaigns can take several forms, each suited to the organization's goals and resources:

- Comprehensive Campaigns are multiyear initiatives designed to raise funds for a wide range of needs: capital projects, operating expenses, endowments, and other programs. These campaigns often integrate other fundraising programs, such as annual giving and planned giving, to contribute to the overall goal. Comprehensive campaigns typically run for three to five years or longer.

- Single-Purpose Campaigns are shorter-term drives focused on funding a single project, such as a building or endowment. These campaigns often target a specific constituent group and may be implemented between the end of one comprehensive campaign and the start of another.

- Sub-Campaigns are smaller initiatives that tie into a continuing major gift program to address minor projects or specific needs (Dove, 2000; Conley, 2016).

Universities, as well as organizations in arts, culture, healthcare, and other sectors, predominantly adopt comprehensive campaigns (Nehls, 2008; Conley, 2016). Comprehensive campaigns may offer greater effectiveness by combining multiple organizational priorities into a single major fundraising effort rather than running separate smaller programs (Newman, 2005).

Phases of a Capital Campaign

Conventionally, capital campaigns are recognized to have two main phases: the quiet phase and the public phase. However, a fully developed campaign typically includes five critical phases (see Figure 1), each contributing to the overall success of the initiative. These phases are:

- Pre-Campaign Planning: This phase involves preparing the organization, assessing potential donors, and creating a comprehensive campaign plan.

- Quiet Phase: During this phase, leadership and major donors are engaged first to secure lead gifts and test the feasibility of the campaign goal.

- Campaign Kick-Off and Public Phase: The campaign is formally launched to the broader community, with events and solicitation efforts designed to expand donor participation.

- Campaign Closing and Celebration: This phase marks the official conclusion of the campaign, providing an opportunity to recognize donors, celebrate successes, and reinforce organizational credibility.

- Stewardship and Campaign Impact: Post-campaign, donors are acknowledged, and the impact of their gifts is communicated, laying the foundation for long-term relationships and future campaigns.

While the first and last phases occur outside the formal campaign period, they are essential. Neglecting these phases can undermine the success of both the current campaign and future fundraising efforts (Conley, 2016). The formal campaign period generally comprises the quiet, public, and closing phases.

These phases provide a structured roadmap for fundraising, ensuring that campaigns progress from preparation and relationship-building to solicitation, celebration, and sustained donor engagement. A deeper exploration of each phase and its activities is provided in Chapters 4 and 5, through the lens of the University of Port Harcourt's maiden campaign.

Figure 1

Phases of a Capital Campaign

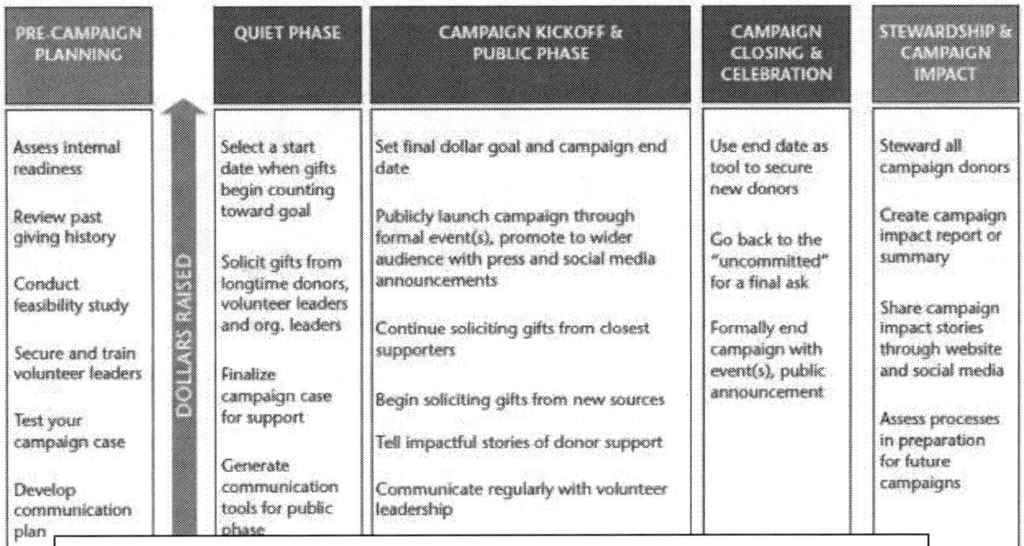

Source: **Figure 19.1** Phases of a Capital Campaign. Reprinted from *Capital campaigns* (p. 246), by A. Conley, 2016, in E. R. Tempel, T. L. Seiler, & D. F. Burlingame (Eds.), *Achieving excellence in fundraising* (4th ed., pp. 243–258). Hoboken, NJ: Wiley. Copyright 2016 by John Wiley & Sons.

Chapter 3

Case Study: The Maiden Campaign – Background and Planning

This chapter presents the background of the University of Port Harcourt's maiden capital campaign and details the full pre-campaign planning process that shaped its launch. In 2006, the University of Port Harcourt launched its first-ever major fundraising effort, aptly titled "Actualizing the Future: Capital Campaign 2006–2010." Spanning five years, the campaign aimed to raise six billion naira, roughly 50 million dollars at the time, to support a broad range of projects: constructing new faculty and administrative buildings, expanding classrooms and hostels, funding student scholarships, establishing new academic departments, and enhancing staff development programs.

The campaign was not just about raising funds; it was a response to the university's urgent need for infrastructure development and financial stability. At the time, the university's development office, known then as the University Advancement Centre (UAC), was still in its infancy, having existed for less than three years. Launching such an ambitious campaign under these conditions was a bold step, reflecting both the institution's aspirations and the emerging profession of university advancement in Nigeria.

Pre-Campaign Planning

Before any fundraising campaign can succeed, careful planning is essential. Whether the program involves annual giving, major gifts, capital campaigns, endowments, or planned giving, effective preparation is key to achieving fundraising goals (Seiler, 2016a). In other words, effective plans must come before effective fundraising.

Pre-campaign planning is a pivotal stage in any fundraising effort. During this phase, all stakeholders, including organizational leadership, the board, fundraisers, and major volunteers, are guided "to begin thinking about the campaign from both the organization's perspective and that of the donors" (Conley, 2016, p. 245). A campaign plan should flow directly from the institution's strategic plan to maintain focus and alignment. It begins with crafting "a compelling case statement" (Conley, 2016, p. 247).

For the University of Port Harcourt's maiden campaign, developing the case for support was the first major task. As Seiler (2016b, p. 44) emphasizes, "The very first step in beginning to develop a case for support originates in the organization's strategic plan." The campaign drew directly from the university's 10-year Strategic Plan 2003–2013, articulating critical needs of the university community that potential donors perceive as valuable (Seiler, 2016a). The case presented persuasive arguments to encourage financial support and addressed the questions donors would naturally ask: why the campaign was vital, how urgent the projects were, and why donors should commit their resources (Ahern, 2009). In this way, the campaign sought not just to raise funds, but to engage donors in the university's vision for its future.

As shown in Figure 1, pre-campaign planning involves several critical steps: assessing internal readiness, reviewing past giving history, conducting a feasibility study, securing and training volunteer leaders, testing the campaign case, and developing a communication plan.

Assessing Internal Readiness

The first pre-campaign planning step is assessing internal readiness. For the University of Port Harcourt's maiden campaign, a formal feasibility study was not conducted. Instead, the development officers relied on extensive training in various aspects of fundraising, including running a capital campaign. The trainings began before the campaign and continued into its first year, provided by institutions such as Pamoja Inc., New York (fundraising consultants with MacArthur Foundation at the time), Council for Advancement and Support of Education (CASE), Community Counseling Service (CCS Fundraising, New York), and the development offices of the University of Washington, Seattle; Jackson State University, Mississippi; and Rensselaer Polytechnic Institute, New York.

The professional experiences and training served as indicators of internal readiness. The development leadership felt confident that the staff could initiate and implement a capital campaign, drawing naturally from the university's strategic plan to identify campaign projects (Kihlstedt, 2010).

Reviewing Past Giving History

Another critical pre-campaign planning step is reviewing past giving history. This involves examining the giving trends of existing donors while researching new prospects, identifying which donor segments have historically supported the institution, and understanding typical gift sizes and frequencies (Seiler, 2016c).

For the University of Port Harcourt's maiden campaign, most potential donors including staff, students, and parents, had no prior history of giving to the university, with the exception of a relatively small number of existing individual, alumni, and corporate donors. Despite this, staff and faculty emerged as substantial sources of support, committing to payroll deduction contributions over a voluntarily stipulated number of years, demonstrating remarkable engagement. Student contributions took the form of a development levy, which differed from the development office's original plan, creating an unresolved question of whether these funds should count toward the campaign goal.

Although the campaign raised significant gifts from only a few donor segments, this step highlights a broader insight: while it is often assumed that donors' resources are limited (Woronkowicz, 2018), organizations may not always tap into all potential funding sources. The effort devoted to identifying and cultivating these sources can greatly influence the campaign's success.

Conducting Feasibility Study

Conducting a feasibility study is particularly important because it "helps an organization determine if they are prepared to undertake the rigor of an ambitious, multiyear campaign" (Conley, 2016, p. 247).

Despite perceived internal readiness, for a development office in its third year, staffed with newly trained personnel, launching a capital campaign was a formidable undertaking. Engaging a consultant for a formal feasibility study would have been ideal, but the emerging profession of university advancement in Nigeria made access to local fundraising consultants difficult. In lieu of a formal feasibility study, the development director and staff created a practical proxy, conducting multiple assessments of known and potential giving sources and refining their strategy.

This process led to the articulation of the campaign theme, the fundraising goal, and the creation of the "Table of Gifts." Modeled similarly to a Gift Range Chart, the Table of Gifts broke down the overall campaign goal by donor segments and specified the number of gifts expected from each. It did not particularly reflect the conventional 80/20 methodology, "where it could be expected that 80 percent of the dollars raised during the campaign will come from just 20 percent of the donors" (Conley, 2016, p. 247), but it was based on careful evaluation of the gift markets and provided a practical roadmap for the campaign.

Evaluating gift markets was a critical component of this step, involving "making informed judgments about which markets to approach and the gift amounts to be sought" (Seiler, 2016a, p. 31). Gifts typically come from a mix of individual donors, corporations, foundations, associations, and government sources. For an educational institution, alumni giving often represents a significant portion of individual contributions.

Securing and Training Volunteer Leaders

Another important step in pre-campaign planning is securing and training volunteer leaders. Volunteers play a key role in fundraising, as they are prepared to engage their peers in support of the campaign. As Seiler (2016a) emphasizes, it is essential to engage volunteers before beginning any fundraising program or campaign, as gift solicitations are most effective when conducted face-to-face by peer volunteers who personally engage potential donors.

In the case of the maiden campaign, however, volunteers were neither engaged nor trained to solicit gifts. This meant that the responsibility for fundraising relied heavily on the development staff and university leadership, highlighting both the challenges and the pioneering nature of launching a capital campaign in an emerging advancement environment.

Volunteer engagement is particularly crucial because the success of a capital campaign often depends on the enthusiastic leadership of both volunteers and institutional executives, and these leaders motivate a committed donor base to make substantial contributions (Bancel, 2000). In the maiden campaign, although there were no major volunteers to lead gift solicitation, the university's leadership stepped into this role. Alongside the development office staff, they conducted a constituency mobilization exercise, visiting different units across campus to solicit support for the campaign. By making their own contributions, the leadership set a positive tone that encouraged other stakeholders to give.

This approach aligns with best practices in fundraising, which recognize the critical role of leadership in successful campaigns. Leaders are expected to act as the chief campaign spokesperson, make personal gifts to the campaign, cultivate donors for major contributions, and actively motivate both staff and volunteers (Nehls, 2008). Through their engagement, the university's executives helped compensate for the absence of formal volunteer leaders and fostered a culture of giving that was essential for the campaign's momentum.

Testing the Campaign Case

Testing the campaign case is another essential step in pre-campaign planning. For the maiden campaign, the case was not tested with external partners. Instead, it was reviewed internally by the university's leadership, who offered their support after participating in sessions led by the development office that explained and educated them about the campaign. The leadership provided valuable input and contributed a foreword and introduction to the campaign document, lending their authority and endorsement to the effort.

However, the development office shared the case for support with other university constituents and potential donors only after the final production of the materials was completed. While this approach ensured a polished presentation, it also limited early engagement and feedback from a broader audience, which could have informed adjustments to enhance donor appeal.

Developing a Communication Plan

Developing a communication plan is another critical step in pre-campaign planning. As Seiler (2016a, p. 33) reminds us, "People give money to causes they know about and care about." A well-crafted communication plan tells the story of the campaign, sharing the case for support in a compelling way that inspires philanthropic engagement.

For the maiden campaign, the development office designed a comprehensive campaign packet, including a brochure with the case statement and additional materials to educate readers about the university and the campaign itself. A key component of this plan was budgeting for communication expenses covering print materials like brochures and newsletters, as well as electronic channels including press coverage, social media, emails, and e-newsletters (Conley, 2016). The plan also accounted for campaign events, ensuring that all aspects of engagement were strategically coordinated.

A fundraising plan of this scope also specifies the programs to be executed, the funding goals for each program, time frames for implementation, the methods to raise the funds, and the responsibilities of staff and volunteers (Seiler, 2016a). Accordingly, the maiden campaign plan outlined a variety of programs and events, including Parents Open Day, Alumni Day, the Major Gift Program, Annual Giving Program, Scholarship Fund Program, the 'Friends of the University dinner, external events, public phase launches in three major cities, and press interviews and releases.

By specifying annual giving, major gifts, and scholarship fund programs separately, the plan recognized that different drives were needed to achieve the overall campaign goal. Gift recognition strategies and stewardship were also included. After months of rigorous work, the final plan was robust, detailing the amount to be raised, projected fundraising expenses, activities to be carried out, targeted donor constituents, solicitation strategies, and the specific projects that would be funded with the donations.

Chapter 4

Case Study: The Maiden Campaign – Implementation and Outcomes

This chapter presents the implementation of the maiden campaign, covering the quiet phase, the public phase, the campaign closing, and the stewardship activities that followed. Once pre-campaign planning was complete, the University of Port Harcourt's maiden campaign moved into the implementation phase. This phase unfolds in three stages: the quiet phase, the public phase, and, ultimately, the campaign closing.

Quiet Phase

The quiet phase, sometimes called the "silent phase," marks the moment "when an organization selects the date in which gifts begin counting toward the campaign goal" (Temple & Seiler, 2016b, p. 249). During this phase, the campaign is known only to leadership and critical actors: board members, major donors, key volunteers, fundraising staff, and other relevant audiences. Typical activities include refining the case for support with input from major donors, soliciting long-standing donors for lead gifts, and finalizing the campaign communication plan. It is also the time when the campaign goal and end date are formally agreed upon.

In many universities, comprehensive campaigns consist of an overall goal composed of smaller, program-specific targets (Conley, 2016), and this was intended for the maiden campaign. Some quiet-phase activities were carried out, but often as part of planning rather than a distinct phase. Notably, input from donors was not sought in refining the case for support, and long-standing donors were not approached for early gifts to test the campaign's goals. The quiet phase was not clearly delineated, and there was no early solicitation to help evaluate whether the campaign goal needed revision or whether the campaign duration should be extended.

Instead, once the campaign materials were ready and the plan appeared complete, the campaign was launched openly, marking the start of the public phase. This approach bypassed the traditional quiet-phase testing and set the stage for broader engagement with the university community and external donors.

Campaign Kick-Off and Public Phase

The public phase of a campaign typically begins with a high-profile event designed to generate excitement and visibility. For the University of Port Harcourt's maiden campaign, this moment arrived in March 2006, during the university's 22nd convocation ceremony.

Ideally, certain steps should be completed just before launching the public phase. One key step is "an assessment of gifts raised to-date and the remaining prospect pool to be solicited in the time left before the campaign's announced end date" (Conley, 2016, p. 250). Best practice suggests that the public phase should begin when roughly 50% of the campaign goal has already been reached. Launching too early, that is, before reaching the halfway mark, may signal a lack of interest and reduce donor confidence, while launching too late, that is, well above 50%, may suggest that the campaign is already successful and dampen the urgency for donations (Conley, 2016).

Campaign solicitation typically follows a top-down and inside-out approach, beginning with "prospective donors at the top of the gift range chart and work down to those in the smaller gift range" (Conley, 2016, p. 250). Moreover, "before the public phase begins, all individuals directly tied to the organization should have made their own campaign commitment," including leadership and staff (Conley, 2016, p. 250; Meyers, 2014).

For the maiden campaign, this approach was only partially followed. Since there was no clearly defined quiet phase, lead gifts from existing major donors were not solicited before the public launch. However, the campaign did begin with internal gifts: the development director and staff pledged monthly payroll deductions over the five-year duration of the campaign, setting a model for soliciting gifts from leadership and staff. Following this, the pro-chancellor, vice-chancellor, and principal officers of the university made their own commitments. Thus, campaign donations were solicited first from university leadership, followed by staff and students, before external stakeholders were approached.

In essence, the university community was educated and mobilized fo the campaign, and their buy-in was secured prior to the public launch However, these internal stakeholders were solicited for actual gifts afte the campaign was publicly launched. This approach highlights both th pioneering efforts and the unique challenges of implementing a capita campaign in a newly established university advancement context.

Campaign Closing and Celebration

The Final Push

"Campaigns come to an end" (Kihlstedt, 2010, p. 7). As a campaig nears its conclusion, it is common practice to widen the pool o potential donors to increase chances of reaching the campaign goal. Thi final round of solicitations often proves more effective when th campaign goal is nearly reached: potential contributors are more easil persuaded to be part of a successful effort. Beyond immediat fundraising, broadening the donor base at this stage strengthen relationships that can benefit future campaigns.

The Role of a Formal Closing

Capital campaigns are typically concluded with "a formal closing even or ceremony to announce the overall dollar amount raised and ke priorities that the campaign helped address" (Conley, 2016, p. 251). Suc events do more than report numbers, they celebrate the connections an relationships forged with donors over the course of the campaign. I this sense, the closing event is as much a relationship milestone as it is a fundraising milestone.

Aligning Campaigns with Leadership Tenure

The maiden campaign at this institution was scheduled for five years, from 2006 to 2010. One key rationale was to align the campaign with the tenure of the university's 6th Vice-Chancellor. This strategy would allow the same leadership to oversee planning, implementation, evaluation, stewardship, and eventual closure. In a context where the university advancement profession was still in early stages of development, neither fully understood nor wholly supported, such continuity was particularly important because a change in leadership could disrupt campaign focus, objectives, and programs.

Challenges in Campaign Closure

While the campaign was launched with a major official event, no formal closing event was held. The Vice-Chancellor completed his tenure in mid-2010, and the development director who had championed the campaign was replaced shortly afterward. Research indicates that the departure of a university president during an ongoing campaign can reduce fundraising momentum and staff morale, placing additional responsibility on the chief development officer and senior staff to maintain continuity (Nehls, 2008).

In this case, the simultaneous exit of both key figures in the campaign's final year contributed directly to the lack of an official closure. Since the Vice-Chancellor's departure was predictable, a closing event could have, and arguably should have, been held before he left office.

Administrative and Operational Hurdles

Following these departures, development staff attempted to ascertain the total funds raised and initiate official campaign closure. These efforts encountered obstacles, partly because the recommendation to track all campaign proceeds in a single designated account had not been followed. Consequently, providing an accurate final total proved challenging, frustrating the development office.

Further complicating matters, handling financial resources fell under the jurisdiction of the bursary. Requests for the bursary to manage campaign funds in a particular way were seen as exceeding the development office's responsibilities. These operational challenges highlighted the growing pains of a nascent university advancement function, an emerging profession still defining its practices and organizational culture.

Stewardship and Campaign Impact

Stewardship is more than accountability; it is "making certain donors' gifts are used wisely, used for intended purposes, managed well in endowments, have an impact, and all this communicated to donors through various ways" (Temple & Seiler, 2016a, p. 432). It is a crucial step toward the next, bigger gift, enhancing donor relations and fostering long-term engagement beyond the act of giving itself. While managing a large campaign donor base can be complex, personal stewardship is strongly recommended. The institution's leadership "should commit the time to personally visit as many campaign donors as possible for the sole purpose of expressing thanks" (Conley, 2016, p. 252). Additional strategies include hosting small group events for different donor categories, distributing impact reports and newsletters in print and electronically, and leveraging the website to keep donors informed. Campaigns can also be memorialized through historic inscriptions that permanently acknowledge "the years of the campaign, the amount raised, and the campaign's impact" (Conley, 2016, p. 252).

In the case of the maiden campaign, the absence of an official closing meant that most of the planned stewardship activities were never implemented, beyond initial thank-you letters. The development officers who planned and executed the campaign repeatedly requested the new leadership to implement appropriate stewardship to honor donors and improve the prospects for future campaigns. Unfortunately, these efforts were met with resistance, and the development staff were therefore reluctant to support a new campaign.

It was not until around 2014, following repeated complaints of the development staff, that an initiative to steward the maiden campaign gifts, carried out by the third development director under the seventh Vice-

Chancellor, was implemented. While the development staff considered this effort modest and belated, it offered an opportunity to recognize contributors in a "Roll of Honor" style document, listing internal stakeholders and other donors alongside a brief summary of the projects supported by the campaign. This printed report was distributed to internal donors. At the end, the original robust stewardship plan for various levels of gifts, including strategies for internal and external donors, was never fully implemented due to internal challenges.

Stewardship, when done well, can be particularly meaningful by tapping into donors' emotional connections. Providing experiences rather than objects, such as facilitating interactions between donors and beneficiaries of their gifts, including scholarship recipients or faculty whose research has been made possible—can evoke lasting emotional engagement and strengthen donor commitment (Ryan, 2018).

Chapter 5
Limitations of the Maiden Campaign

The preceding analysis highlights that the maiden campaign had notable achievements, but it also revealed several critical limitations discussed below. These limitations, when carefully considered, provide valuable insights for future campaigns.

Limitations

Board, Volunteers, and Development Committee

The first and perhaps most critical limitation was the absence of a board of trustees or directors, no volunteers, and no development committee at the time the maiden campaign was planned and implemented. The campaign had a lofty goal of 6 billion naira. University leadership provided financial resources to run the campaign, and the development office had trained and competent staff. On the surface, it seemed there were adequate resources to embark on the ambitious journey of the university's first capital campaign, the first in the history of University Advancement in the country.

What was missing, however, was critical: leadership engagement through a board and volunteers. As Henry A. Rosso aptly stated, "Fundraising begins with the board" (Temple & Seiler, 2016b, p. 439), and this principle is even more crucial for campaigns pursuing audacious goals.

Volunteers, often major donors and distinguished alumni capable of influencing the organization and its fundraising (Seiler, 2016a), were not involved, which inevitably limited the campaign's reach and impact. Similarly, the planned development committee was never constituted. The lack of these key actors constrained the development office's ability to tap into the enormous potential funding sources available. In effect, while the campaign goal was substantial, it remained modest relative to the financial resources the university could have accessed with broader support.

Feasibility Study

A second limitation was the absence of a formal feasibility study. Such a study not only projects a campaign's financial goal but also provides "a realistic assessment of the organization's potential donor base and whether enough prospective donors exist to reach the goal" (Conley, 2016, p. 247). The development office's internal assessment of readiness and external stakeholders' capacity and willingness to support the campaign could be argued as incomplete, since consultations with key external stakeholders and existing major donors were not undertaken.

Constituency Mobilization Timing

Third, the campaign brochure was printed before a comprehensive constituency mobilization across the campus. This mobilization was meant to educate staff and students about the maiden campaign and persuade them to own the case for support. Instead, the internal constituents felt their input was not sought in developing the campaign case, generating friction between the development office, university leadership, and the university community. Resolving this required additional meetings and outreach by the development office with key staff and student unions, demonstrating the importance of early engagement in campaign planning.

Program Implementation and Resource Utilization

Fourth, several key events and programs planned for the maiden campaign were never implemented, leaving mapped financial resources underutilized. Had all planned programs been executed, the campaign might have raised even greater support, as early events had already generated substantial gifts, such as the successful launch of the campaign, a high-profile alumni event in the Federal Capital Territory featuring an alumnus who was serving as a state governor, and the launch of the annual giving program. Possible reasons for incomplete implementation include limited human resources, particularly a lack of volunteers to lead gift solicitation, and efforts to solicit gifts from sources where the university had no connections, which may have led to staff burnout. Additionally, early receipts of large gifts may have created a sense of premature campaign success, reducing urgency to pursue further donations.

Timing of the Annual Giving Program

Fifth, the Annual Giving program, which is a foundational steppingstone for major gifts, was not operational when the campaign began. It was launched three years into the campaign. A robust annual fund donor base is critical for building reliable major gift programs, and in this case the sequencing of fundraising initiatives was largely reversed.

Nature of Campaign Closing

Sixth, the maiden campaign was never officially closed. While the campaign had a high-profile opening, no closing ceremony was held to announce the total amount raised. Judging campaign success solely by funds raised becomes difficult in such cases, especially since campaign success is traditionally defined as meeting or exceeding the set goal. The development office estimated that the goal was likely reached or even surpassed over time, but without formal closure, this success could not be publicly recognized or celebrated.

Stewardship of Gifts

Finally, stewardship, which is the process of ensuring that gifts are used wisely and acknowledged properly, was delayed and incomplete. Although initial thank-you letters were sent, the robust stewardship plan envisioned during campaign planning was largely unrealized. Proper and timely stewardship not only fosters donor satisfaction but also lays the groundwork for future contributions.

Taken together, the limitations of the maiden campaign illuminate the challenges faced by a nascent university advancement office. They underscore the importance of leadership, volunteer engagement, thorough planning, stakeholder consultation, and timely execution of campaign milestones for future fundraising success.

Chapter 6

Recommendations for Future Capital Campaigns

Reflecting on the limitations of the maiden campaign, several key insights emerge that are instructive for future university fundraising efforts. The following recommendations are proposed to guide future capital campaigns at the University of Port Harcourt and any advancement office embarking on a major fundraising initiative.

Pre-Campaign Planning Best Practices

1. Establish Leadership Structures and Volunteer Base

The presence of a board, development committee, and volunteers is foundational for campaign success. Leadership provides credibility and opens doors to networks, potential major donors, and organizational influence. Before launching a capital campaign, the development office or office of institutional advancement should have a board of trustees or directors in place, constitute a development committee, and recruit and train volunteers.

As Temple and Seiler (2016, p. 441) note, "Board members must ask questions that ensure that the fundraising strategies and tactics employed by the organization are based on reliable data, good intuition, and sound practices." Volunteers play a key role in major gift solicitation and foster emotional connections with donors, as "donors give to people and emotional needs, not things and numerical needs" (Ryan, 2018, p. 44). Without these actors, even a well-trained development office may struggle to mobilize sufficient resources.

2. Conduct Feasibility Studies

A formal feasibility study is critical for realistic goal-setting. Where hiring an external consultant is not feasible, a proxy assessment by key volunteers and major donors can evaluate giving history and expand the donor base prior to the campaign. Understanding the capacity and willingness of potential donors helps set achievable goals, prioritize gift markets, and tailor solicitation strategies effectively.

3. Early Stakeholder Engagement

Constituency mobilization should precede the finalization of campaign materials. Key stakeholders, including university leadership, faculty, staff, students, and external supporters, should be involved early in the planning process. Soliciting their opinions helps identify priority projects and informs the case for support. Integrating stakeholder feedback ensures the campaign addresses what donors care about, increasing the likelihood of financial commitment and fostering a sense of ownership.

4. Comprehensive Campaign Plan

The campaign plan should detail all planned events, cultivation strategies and solicitation activities. Planning alone is insufficient; campaigns succeed only when planned programs are fully executed. Strategic sequencing of fundraising initiatives is essential: launching an annua giving program first to build a broad donor base, followed by a major gift program to cultivate lead donors, before initiating a full capita campaign. Each stage strengthens the pipeline of donors and positions the organization to secure critical lead gifts.

Implementation and Solicitation Strategies

1. Clearly Defined Campaign Phases

A capital campaign should include distinct quiet, public, and closing phases. The quiet phase allows solicitation of lead gifts and testing of campaign goals. The public phase follows with a high-profile launch, and the closing phase celebrates achievements, reinforces stewardship, and provides a natural pause before planning future campaigns. Structured campaign phases support strategic solicitation, maintain dono momentum, and enhance overall campaign effectiveness.

2. Strategic Solicitation

Campaign staff and volunteers should follow a structured approach engaging leadership and top prospective donors first, then moving to smaller gift levels. Resource planning must consider staff and voluntee capacity. Misalignment between campaign goals and human resources can lead to staff burnout, incomplete program implementation, and lost opportunities.

Closing and Stewardship Best Practices

1. Timely Campaign Closure

Officially closing a campaign is as important as launching it. The closing event provides a platform to celebrate success, recognize donors, and reinforce trust. Formal closure also facilitates future campaigns by demonstrating professionalism and accountability.

2. Robust Stewardship

Donor stewardship is a relationship-building exercise, not merely an administrative task. It should begin immediately upon receipt of gifts and continue throughout and beyond the campaign. Effective stewardship entails personal engagement with donors, acknowledgment of contributions, sharing the impact of gifts, and providing meaningful opportunities for involvement. Neglecting donor stewardship can weaken relationships and undermine future campaigns, but consistent and thoughtful engagement can strengthen connections with donors, enhancing the outcomes of current and future campaigns (Ryan, 2018).

Building a Sustainable Donor Culture

1. Evaluate Gift Markets and Potential Donors

Pre-campaign planning should include a thorough assessment of gift markets and identification of potential donors. Accurate evaluation informs realistic campaign goals and enhances the likelihood of success.

2. Campaigns as Organizational Transformation

Beyond financial goals, capital campaigns serve as catalysts for institutional transformation. Major campaigns help organizations focus on vision and values, fostering lasting cultural change (O'Brien, 1998).

These recommendations collectively underscore that capital campaigns are complex undertakings requiring more than financial resources. Success relies on careful planning, strategic leadership, stakeholder engagement, phased execution, and a focus on relationship-building. For nascent university advancement offices, the maiden campaign serves as both a cautionary tale and a rich source of guidance for designing more effective campaigns in the future.

Chapter 7

Conclusion – Reflections and Future Directions

The maiden capital campaign at the University of Port Harcourt offers valuable lessons that extend well beyond the funds raised. While the campaign faced significant limitations in planning and implementation, it catalyzed infrastructural transformation, introduced capital campaign practices to the university, and began fostering a culture of philanthropy. This chapter reflects on the broader value of the campaign, the evolution of the University Advancement Office into a foundation, and the implications for future capital campaigns in Nigeria.

Reflections on the Value of Campaigns Beyond Fundraising

The maiden campaign at the University of Port Harcourt, despite its limitations, demonstrated that the value of a capital campaign extends well beyond the amount of funds raised. While judged a success in terms of financial contributions, the most significant outcomes were infrastructural improvements and institutional transformation. The campaign galvanized the university community, introduced capital campaign practices, and began establishing a culture of philanthropy.

The campaign illustrated that planning itself constitutes a significant achievement. Organizing a major fundraising initiative without a board, volunteers, or an established culture of giving was a daunting endeavor. In this sense, the maiden campaign achieved successes independent of its fundraising results. As Kihlstedt and Pierpont (1998, p. 10) note, a capital campaign can "have a galvanizing impact and value well beyond the resources it will engender."

During the campaign, terms such as "Capital Campaign," "Fundraising," and "Development Officers" became widely recognized across the university community. Engagement extended to staff, students, alumni, and corporate supporters. Students initially resistant to the campaign were converted, and student leadership ultimately contributed to its success. The campaign fostered awareness, ownership, and a collective commitment to the university's development mission.

The Evolution of the University Advancement Office

A key outcome of the maiden campaign was the evolution of the University Advancement Office (UAC). Originally operating with only ten staff members, the office has since transitioned into the University of Port Harcourt Foundation, complete with a board of trustees and an alumni advisory board. Some faculties and departments began to establish their own fundraising units, thereby providing a more sustainable infrastructure for institutional advancement.

This evolution demonstrates that even in the absence of ideal structures such as a board, volunteers, and a development committee, the resilience and professionalism of staff, combined with the cultural context of giving, can produce remarkable results. Nonetheless, the experience underscores the importance of adhering to established fundraising principles to cultivate sustained donor engagement and avoid one-off donations.

Vision for Future Campaigns in Nigeria

Looking forward, future capital campaigns at the University of Port Harcourt and across Nigerian higher education institutions, should build on the lessons learned from the maiden campaign. Effective planning, early stakeholder engagement, structured campaign phases, and robust stewardship practices are essential for maximizing both fundraising outcomes and institutional impact.

Future campaigns are expected to:

- Set more realistic goals aligned with donor capacity.
- Implement comprehensive pre-campaign planning, including volunteer recruitment and board engagement.
- Execute events, cultivation strategies, and solicitation activities fully and strategically.

- Ensure timely stewardship and ongoing engagement to transform one-time donors into sustained supporters.

The maiden campaign also highlights the potential for campaigns to act as catalysts for organizational transformation. Beyond resources, campaigns focus attention on institutional vision and values, fostering cultural change and enhancing the university's profile.

Finally, the dedication and resilience of the development staff during the maiden campaign merit recognition. Their efforts laid a foundation not only for the University of Port Harcourt's ongoing advancement work but also for the broader professionalization of fundraising in Nigerian higher education. Future campaigns will continue this trajectory, demonstrating that well-planned and executed capital campaigns can achieve lasting impact, both financially and institutionally.

References

Ahern, T. (2009). *Seeing through a donor's eyes: How to make a persuasive case for everything from your annual drive to your planned giving program to your capital campaign.* Emerson & Church.

Bancel, M. (2000). *Preparing Your Capital Campaign.* Excellence in Fund Raising Workbook Series. San Francisco: Jossey-Bass.

Conley, A. (2016). Capital campaigns. In E. R. Tempel, T. L. Seiler, & D. F. Burlingame (Eds.), *Achieving excellence in fundraising* (4th ed., pp. 243–258). Wiley.

Dove, K. E. (2000). *Conducting a successful capital campaign* (2nd ed.). Jossey-Bass.

Kihlstedt, A. (2010). *Capital campaigns: Strategies that work* (3rd ed.). Jones and Bartlett.

Meyers, H. (2014, November 4). Capital campaign strategies you may be missing. *University Business.* https://universitybusiness.com/capital-campaign-strategies-you-may-be-missing/

Nehls, K. (2008). Presidential transitions during capital campaigns. *International Journal of Educational Advancement, 8*(3–4), 198–218. https://doi.org/10.1057/ijea.2009.3

Newman, D. S. (2005). The risk of capital-campaign fatigue. *Chronic of Philanthropy, 17*(20), B22–B24.

O'Brien, C. L. (1998). Thinking beyond the dollar goal: A campaig as organizational transformation. In A. Kihlstedt & R. Pierpon (Eds.), *Capital campaigns: Realizing their power and potential* (Ne Directions for Philanthropic Fundraising, No. 21). Jossey-Bass.

Rosso, H. A. (2016). A philosophy of fundraising. In E. R. Tempe T. L. Seiler, & D. F. Burlingame (Eds.), *Achieving excellence in fundraisin* (4th ed., pp. 3–10). Wiley.

Ryan, E. (2018, July/August). Recharge your campaign: The formul for energizing your donors = excitement + emotion. *Currents, 44*(4) 40-46.

Seiler, T. L. (2016a). Plan to succeed. In Tempel, E. R., Seiler, T. L & Burlingame, D. F. (Eds.), *Achieving excellence in fundraising* (4th ed pp. 27–35). Wiley.

Seiler, T. L. (2016b). Developing and articulating a case for support In Tempel, E. R., Seiler, T. L., & Burlingame, D. F. (Eds.), *Achievin excellence in fundraising* (4th ed., pp. 37–48). Wiley.

Seiler, T. L. (2016c). Individuals as a constituency for fundraising. I Tempel, E. R., Seiler, T. L., & Burlingame, D. F. (Eds.), *Achievin excellence in fundraising* (4th ed., pp. 49–58). Wiley.

Sevier, R. A. (2003, May). Marketing support for capital campaigns: Get your story across with solid marketing strategy and tactics, but never forget the people. *University Business, 6*(5), 44-47.

Temple, E. R., & Seiler, T. L. (2016a). Stewardship and accountability. In Tempel, E. R., Seiler, T. L., & Burlingame, D. F. (Eds.), *Achieving excellence in fundraising* (4th ed., pp. 431–438). Wiley.

Temple, E. R., & Seiler, T. L. (2016b). Engaging the board in fundraising. In Tempel, E. R., Seiler, T. L., & Burlingame, D. F. (Eds.), *Achieving excellence in fundraising* (4th ed., pp. 439–450). Wiley.

The Fund Raising School. (2017). *Principles and techniques of fundraising.* Indiana University Lilly Family School of Philanthropy.

Washington, B. T. (1901). *Raising money.* In *Up from slavery: An autobiography.*

Woronkowicz, J. (2018). The effects of capital campaigns on local nonprofit ecologies. *Nonprofit and Voluntary Sector Quarterly, 47*(3), 645–656. https://doi.org/10.1177/0899764018757026

Made in the USA
Monee, IL
05 November 2026

68065366R00028